See It Grow

BANANA

by Thomas Persano

Consultant: Karen C. Hall, PhD
Botanist
Eugene, Oregon

BEARPORT
PUBLISHING

New York, New York

Credits

Cover, © Valentyn Volkov/Alamy; Title Page, © BookyBuggy/Shutterstock; TOC, © panda3800/ Shutterstock and © Cbenjasuwan/Shutterstock; 4–5, © Valentyn75/Dreamstime; 5R, © Oksana Kuzmina/Shutterstock; 6–7, © ImpakPro/iStock; 8, © gan chaonan/Shutterstock and © Obsidian Soul/CCO; 9T, © gan chaonan/Shutterstock, © Obsidian Soul/CCO, and © Africa Studio/Shutterstock; 9B, © Davdeka/Shutterstock; 10, © natthawut ngoensanthia/Shutterstock; 11, © Muangsatun/ Shutterstock; 12T, © JosephPaul213/Shutterstock; 12B, © Lamyai/Shutterstock; 13, © szefei/ Shutterstock; 14, © Iryna Rasko/Alamy; 15, © Markuso53/Dreamstime; 16, © oradige59/ Shutterstock; 17, © luis2499/Shutterstock; 18, © denizuniusu/iStock; 19, © MIA Studio/Shutterstock; 20 (T to B), © Olha Huchek/Dreamstime, © Tevarak/iStock, and © Misko Kordic/Dreamstime; 21 (T to B), © naluwan/Shutterstock, © Westernd61/Alamy, and © Bidouze Stéphane/Dreamstime; 22, © Mkumaresa/CC BY-SA 4.0; 23 (T to B), © TS Photographer/Shutterstock, © garmoncheg/ Shutterstock, © barmalini/Shutterstock, © gan chaonan/Shutterstock, © Obsidian Soul/CC0, © Africa Studio/Shutterstock, and © Davdeka/Shutterstock; 24, © Olexandr Panchenko/Shutterstock; Back Cover, © Hanna_photo/Shutterstock.

Publisher: Kenn Goin
Senior Editor: Joyce Tavolacci
Creative Director: Spencer Brinker
Design: Debrah Kaiser
Photo Researcher: Thomas Persano

Library of Congress Cataloging-in-Publication Data

Names: Persano, Thomas, author.
Title: Banana / by Thomas Persano.
Description: New York, New York: Bearport Publishing Company, 2019. |
Series: See it grow | Includes bibliographical references and index.
Identifiers: LCCN 2018009264 (print) | LCCN 2018011403 (ebook) |
ISBN 9781684027170 (ebook) | ISBN 9781684026715 (library)
Subjects: LCSH: Bananas—Juvenile literature.
Classification: LCC SB379.B2 (ebook) | LCC SB379.B2 P54 2018 (print) |
DDC 634/.772—dc23
LC record available at https://lccn.loc.gov/2018009264

For more information, write to Bearport Publishing Company, Inc., 45 West 21st Street, Suite 3B, New York, New York 10010. Printed in the United States of America.

10 9 8 7 6 5 4 3 2 1

Contents

Banana

Bananas are long, with a tough, yellow peel.

Inside, they are soft and sweet.

How did they get that way?

Bananas are a popular breakfast food!

Banana plants grow in warm, sunny places.

They look like trees, but they are **herbs**.

In fact, they're the largest herbs in the world!

A banana plant can grow up to 25 feet (7.6 m) tall.

A banana plant grows from an underground stem.

It's called a rhizome (RYE-zohm).

The rhizome is covered with **roots.**

Other plants grow from rhizomes, too. These include ginger and asparagus.

A small, green **stalk** pushes out of the rhizome.

This is a baby banana plant.

The stalk is called a pseudostem (SOO-doh-stehm).

The pseudostem is made up of many folded leaves.

Over time, the pseudostem gets bigger.

Larger and larger leaves grow out of it.

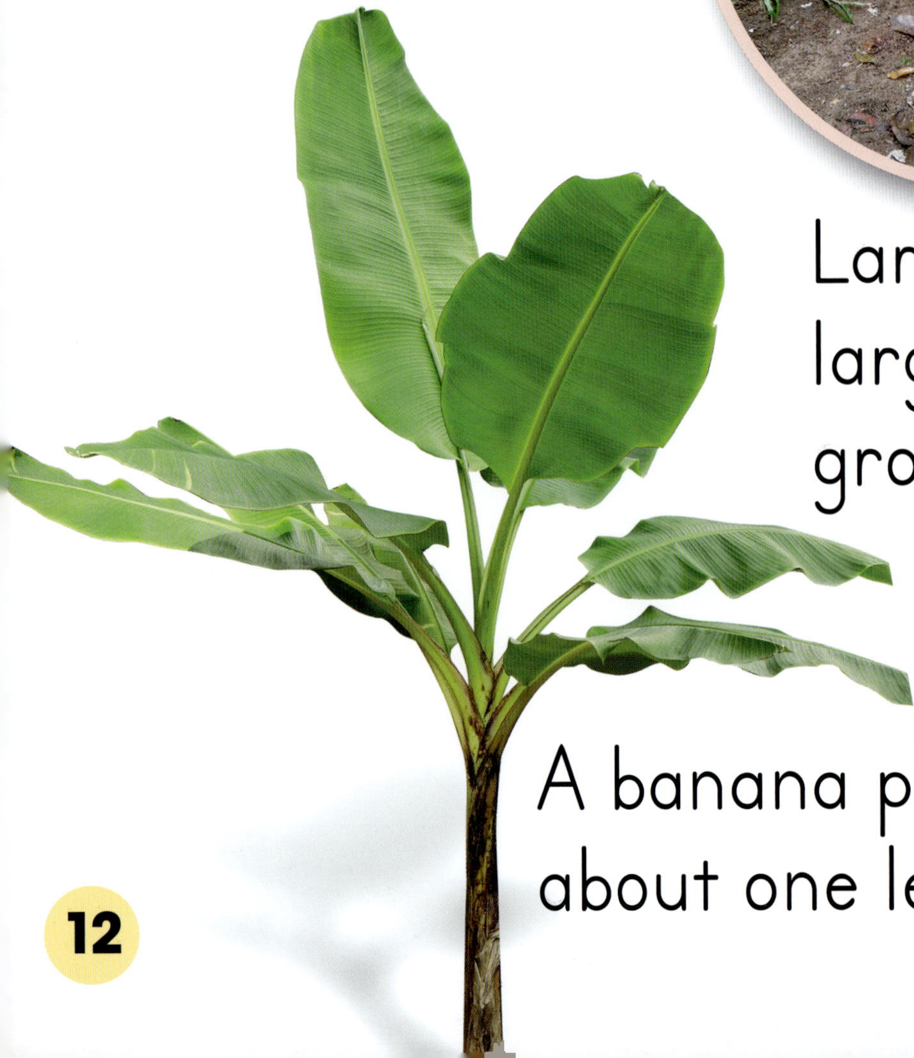

A banana plant grows about one leaf per week.

In some countries, banana leaves are used as plates!

The plant grows another stem.

It's long and thin.

A huge purple **bud** forms at the end of it.

The bud forms when a banana plant is about a year old.

The bud slowly opens.

Rows of little flowers appear.

Some of the flowers grow into tiny fruits.

These are baby bananas!

Groups of bananas are called hands. Each banana is called a finger!

The small bananas grow and grow.

After about 100 days, they're ready to be picked.

Bananas curve upward as they grow!

Most bananas are picked when they're green.

As they **ripen**, they turn yellow.

That's when they're sweet and ready to eat. Yum!

There are many kinds of bananas. Some are red, brown, or purple.

Banana Facts

- Most of the world's bananas are grown in the Americas.
- Billions of bananas are eaten each year.
- People have been growing bananas for over 7,000 years.
- The most popular type of banana is called the Cavendish.
- Some wild bananas have lots of hard seeds.

Glossary

bud (BUHD) a growth on the stem of a plant that turns into a leaf or flower

herbs (URBZ) plants that don't have woody stems

ripen (RYE-pihn) to become fully grown and ready for eating

roots (ROOTS) parts of a plant that take in water and nutrients from the soil

stalk (STAWK) a long, thin part of a plant

Index

Read More

Rattini, Kristin Baird. *Seed to Plant.* Washington, DC: National Geographic (2014).

Williams, Kathryn. *Plants.* Washington, DC: National Geographic (2017).

Learn More Online

To learn more about bananas, visit
www.bearportpublishing.com/SeeItGrow

About the Author

Thomas Persano lives in New York City. When he was a kid, he hated bananas. Now, he eats them all the time!